THE COMMON CORE

Clarifying Expectations for Teachers & Students

MATH

Grade 8

Created and Presented by
Align, Assess, Achieve

Education

Align, Assess, Achieve, LLC

STEM McGraw-Hill is committed to providing instructional materials in Science, Technology, Engineering, and Mathematics (STEM) that give all students a solid foundation, one that prepares them for college and careers in the 21st century.

Send all inquiries to:
McGraw-Hill Education
STEM Learning Solutions Center
8787 Orion Place
Columbus, OH 43240

ISBN: 978-007-662900-8
MHID: 0-07-662900-7

Printed in the United States of America.

1 2 3 4 5 6 7 8 9 QLM 16 15 14 13 12 11

STEM

Our mission is to provide educational resources that enable students to become the problem solvers of the 21st century and inspire them to explore careers within Science, Technology, Engineering, and Mathematics (STEM) related fields.

The *McGraw·Hill* Companies

Acknowledgements

This book integrates the Common Core State Standards – a framework for educating students to be competitive at an international level – with well-researched instructional planning strategies for achieving the goals of the CCSS. Our work is rooted in the thinking of brilliant educators, such as Grant Wiggins, Jay McTighe, and Rick Stiggins, and enriched by our work with a great number of inspiring teachers, administrators, and parents. We hope this book provides a meaningful contribution to the ongoing conversation around educating lifelong, passionate learners.

We would like to thank many talented contributors who helped create *The Common Core: Clarifying Expectations for Teachers and Students.* Our author Laura Hance, for her intelligence, persistence, and love of teaching; Graphic Designer Thomas Davis, for his creative talents and good nature through many trials; Editors, Sandra Baker, Dr. Teresa Dempsey, and Wesley Yuu, for their educational insight and deep understanding of mathematics; Director of book editing and production Josh Steskal, for his feedback, organization, and unwavering patience; Our spouses, Andrew Bainbridge and Tawnya Holman, who believe in our mission and have, through their unconditional support and love, encouraged us to take risks and grow.

Katy Bainbridge
Bob Holman
Co-Founders
Align, Assess, Achieve, LLC

Executive Editors: *Katy Bainbridge, Bob Holman, Sandra Baker, and Wesley Yuu*
Contributing Authors: *Deborah L. Kaiser, Theresa Mariea, Laura Hance, Ali Fleming, Melissa L. McCreary, Charles L. Brads, Teresa Dempsey, Rebecca Watkins-Heinze, Bob Holman, Wesley Yuu*
Editors: *Jason Bates, Charles L. Brads, Marisa Hilvert, Stephanie Archer*
Graphic Design & Layout: *Thomas Davis; thomasanceldesign.com*
Director of Book Editing & Production: *Josh Steskal*

Introduction

Purpose

The Common Core State Standards (CCSS) provide educators across the nation with a shared vision for student achievement. They also provide a shared challenge: how to interpret the standards and use them in a meaningful way? Clarifying the Common Core was designed to facilitate the transition to the CCSS at the district, building, and classroom level.

Organization

Clarifying the Common Core presents content from two sources: the CCSS and Align, Assess, Achieve. Content from the CCSS is located in the top section of each page and includes the domain, cluster, and grade level standard. The second section of each page contains content created by Align, Assess, Achieve – Enduring Understandings, Essential Questions, Suggested Learning Targets, and Vocabulary. The black bar at the bottom of the page contains the CCSS standard identifier. A sample page can be found in the next section.

Planning for Instruction and Assessment

This book was created to foster meaningful instruction of the CCSS. This requires planning both quality instruction and assessment. Designing and using quality assessments is key to high-quality instruction (Stiggins et al.). Assessment should accurately measure the intended learning and should inform further instruction. This is only possible when teachers and students have a clear vision of the intended learning. When planning instruction it helps to ask two questions, "Where am I taking my students?" and "How will we get there?" The first question refers to the big picture and is addressed with **Enduring Understandings** and **Essential Questions**. The second question points to the instructional process and is addressed by **Learning Targets**.

Where Am I Taking My Students?

When planning, it is useful to think about the larger, lasting instructional concepts as **Enduring Understandings**. Enduring Understandings are rooted in multiple units of instruction throughout the year and are often utilized K-12. These concepts represent the lasting understandings that transcend your content. Enduring Understandings serve as the ultimate goal of a teacher's instructional planning. Although tempting to share with students initially, we do not recommend telling students the Enduring Understanding in advance. Rather, Enduring Understandings are developed through meaningful engagement with an Essential Question.

(continued on next page)

Essential Questions work in concert with Enduring Understandings to ignite student curiosity. These questions help students delve deeper and make connections between the concepts and the content they are learning. Essential Questions are designed with the student in mind and do not have an easy answer; rather, they are used to spark inquiry into the deeper meanings (Wiggins and McTighe). Therefore, we advocate frequent use of Essential Questions with students. It is sometimes helpful to think of the Enduring Understanding as the answer to the Essential Question.

How Will We Get There?

If Enduring Understandings and Essential Questions represent the larger, conceptual ideas, then what guides the learning of specific knowledge, reasoning, and skills? These are achieved by using **Learning Targets**. Learning Targets represent a logical, student friendly progression of teaching and learning. Targets are the scaffolding students climb as they progress towards deeper meaning.

There are four types of learning targets, based on what students are asked to do: knowledge, reasoning/understanding, skill, and product (Stiggins et al.). When selecting Learning Targets, teachers need to ask, "What is the goal of instruction?" After answering this question, select the learning target or targets that align to the instructional goal.

Instructional Goal	Types of Learning Targets	Key Verbs
Recall basic information and facts	Knowledge (K)	Name, identify, describe
Think and develop an understanding	Reasoning/Understanding (R)	Explain, compare and contrast, predict
Apply knowledge and reasoning	Skill (S)	Use, solve, calculate
Synthesize to create original work	Product (P)	Create, write, present

Adapted from Stiggins et al. *Classroom Assessment for Student Learning.* (Portland: ETS, 2006). Print.

Each book contains two types of Enduring Understandings and Essential Questions. The first type, located on the inside cover, relate to the Mathematical Practices, which apply K-12. The second type are based on the domain, cluster, and standard and are located beneath each standard.

Keep in mind that the Enduring Understandings, Essential Questions, and Learning Targets in this book are suggestions. Modify and combine the content as necessary to meet your instructional needs. Quality instruction consists of clear expectations, ongoing assessment, and effective feedback. Taken together, these promote meaningful instruction that facilitates student mastery of the Common Core State Standards.

References

Stiggins, Rick, Jan Chappuis, Judy Arter, and Steve Chappuis. *Classroom Assessment for Student Learning.* 2nd. Portland, OR: ETS, 2006.

Wiggins, Grant, and Jay McTighe. *Understanding by Design, Expanded 2nd Edition.* 2nd. Alexandria, VA: ASCD, 2005.

Page Organization

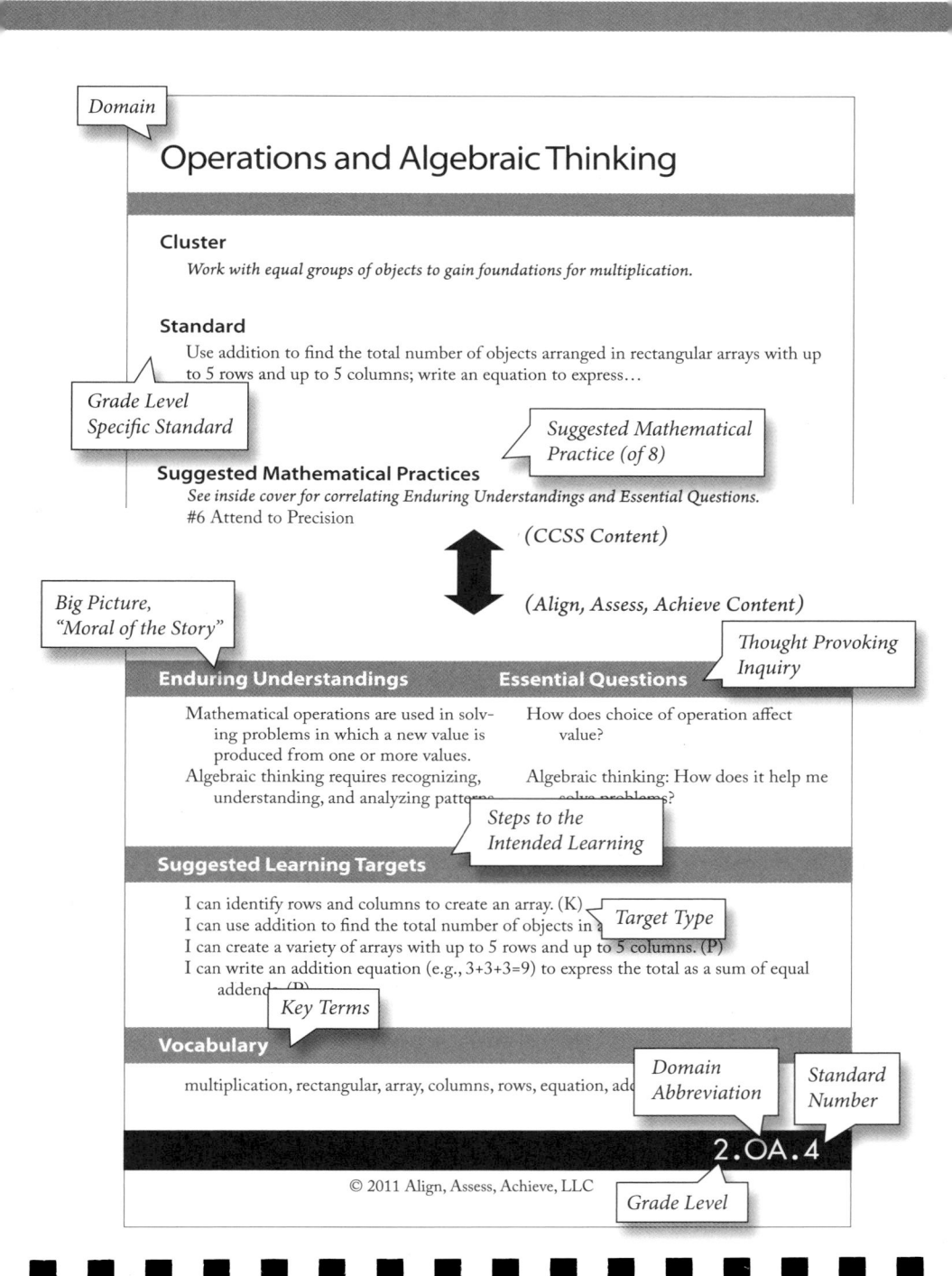

Domain

Operations and Algebraic Thinking

Cluster

Work with equal groups of objects to gain foundations for multiplication.

Standard

Use addition to find the total number of objects arranged in rectangular arrays with up to 5 rows and up to 5 columns; write an equation to express…

Grade Level Specific Standard

Suggested Mathematical Practice (of 8)

Suggested Mathematical Practices

See inside cover for correlating Enduring Understandings and Essential Questions.
#6 Attend to Precision

(CCSS Content)

(Align, Assess, Achieve Content)

Big Picture, "Moral of the Story"

Thought Provoking Inquiry

Enduring Understandings

Mathematical operations are used in solving problems in which a new value is produced from one or more values.
Algebraic thinking requires recognizing, understanding, and analyzing patterns.

Essential Questions

How does choice of operation affect value?

Algebraic thinking: How does it help me solve problems?

Steps to the Intended Learning

Suggested Learning Targets

I can identify rows and columns to create an array. (K)
I can use addition to find the total number of objects in
I can create a variety of arrays with up to 5 rows and up to 5 columns. (P)
I can write an addition equation (e.g., 3+3+3=9) to express the total as a sum of equal addends. (P)

Target Type

Key Terms

Vocabulary

multiplication, rectangular, array, columns, rows, equation, add

Domain Abbreviation

Standard Number

2.OA.4

© 2011 Align, Assess, Achieve, LLC

Grade Level

Mathematical Practices

#1 Making sense of problems and persevere in solving them.

Mathematically proficient students start by explaining to themselves the meaning of a problem and looking for entry points to its solution. They analyze givens, constraints, relationships, and goals. They make conjectures about the form and meaning of the solution and plan a solution pathway rather than simply jumping into a solution attempt. They consider analogous problems, and try special cases and simpler forms of the original problem in order to gain insight into its solution. They monitor and evaluate their progress and change course if necessary. Older students might, depending on the context of the problem, transform algebraic expressions or change the viewing window on their graphing calculator to get the information they need. Mathematically proficient students can explain correspondences between equations, verbal descriptions, tables, and graphs or draw diagrams of important features and relationships, graph data, and search for regularity or trends. Younger students might rely on using concrete objects or pictures to help conceptualize and solve a problem. Mathematically proficient students check their answers to problems using a different method, and they continually ask themselves, "Does this make sense?" They can understand the approaches of others to solving complex problems and identify correspondences between different approaches.

Mathematically proficient students make sense of quantities and their relationships in problem situations. They bring two complementary abilities to bear on problems involving quantitative relationships: the ability to *decontextualize*—to abstract a given situation and represent it symbolically and manipulate the representing symbols as if they have a life of their own, without necessarily attending to their referents—and the ability to *contextualize*, to pause as needed during the manipulation process in order to probe into the referents for the symbols involved. Quantitative reasoning entails habits of creating a coherent representation of the problem at hand; considering the units involved; attending to the meaning of quantities, not just how to compute them; and knowing and flexibly using different properties of operations and objects.

#3. Construct viable arguments and critique the reasoning of others.

Mathematically proficient students understand and use stated assumptions, definitions, and previously established results in constructing arguments. They make conjectures and build a logical progression of statements to explore the truth of their conjectures. They are able to analyze situations by breaking them into cases, and can recognize and use counterexamples. They justify their conclusions, communicate them to others, and respond to the arguments of others. They reason inductively about data, making plausible arguments that take into account the context from which the data arose. Mathematically proficient students are also

(continued on next page)

able to compare the effectiveness of two plausible arguments, distinguish correct logic or reasoning from that which is flawed, and—if there is a flaw in an argument—explain what it is. Elementary students can construct arguments using concrete referents such as objects, drawings, diagrams, and actions. Such arguments can make sense and be correct, even though they are not generalized or made formal until later grades. Later, students learn to determine domains to which an argument applies. Students at all grades can listen or read the arguments of others, decide whether they make sense, and ask useful questions to clarify or improve the arguments.

#4 Model with mathematics.

Mathematically proficient students can apply the mathematics they know to solve problems arising in everyday life, society, and the workplace. In early grades, this might be as simple as writing an addition equation to describe a situation. In middle grades, a student might apply proportional reasoning to plan a school event or analyze a problem in the community. By high school, a student might use geometry to solve a design problem or use a function to describe how one quantity of interest depends on another. Mathematically proficient students who can apply what they know are comfortable making assumptions and approximations to simplify a complicated situation, realizing that these may need revision later. They are able to identify important quantities in a practical situation and map their relationships using such tools as diagrams, two-way tables, graphs, flowcharts and formulas. They can analyze those relationships mathematically to draw conclusions. They routinely interpret their mathematical results in the context of the situation and reflect on whether the results make sense, possibly improving the model if it has not served its purpose.

#5 Use appropriate tools strategically.

Mathematically proficient students consider the available tools when solving a mathematical problem. These tools might include pencil and paper, concrete models, a ruler, a protractor, a calculator, a spreadsheet, a computer algebra system, a statistical package, or dynamic geometry software. Proficient students are sufficiently familiar with tools appropriate for their grade or course to make sound decisions about when each of these tools might be helpful, recognizing both the insight to be gained and their limitations. For example, mathematically proficient high school students analyze graphs of functions and solutions generated using a graphing calculator. They detect possible errors by strategically using estimation and other mathematical knowledge. When making mathematical models, they know that technology can enable them to visualize the results of varying assumptions, explore consequences, and compare predictions with data. Mathematically proficient students at various grade levels are able to identify relevant external mathematical resources, such as digital content located on a website, and use them to pose or solve problems. They are able to use technological tools to explore and deepen their understanding of concepts.

(continued on next page)

#6 Attend to precision.

Mathematically proficient students try to communicate precisely to others. They try to use clear definitions in discussion with others and in their own reasoning. They state the meaning of the symbols they choose, including using the equal sign consistently and appropriately. They are careful about specifying units of measure, and labeling axes to clarify the correspondence with quantities in a problem. They calculate accurately and efficiently, express numerical answers with a degree of precision appropriate for the problem context. In the elementary grades, students give carefully formulated explanations to each other. By the time they reach high school they have learned to examine claims and make explicit use of definitions.

#7 Look for and make use of structure.

Mathematically proficient students look closely to discern a pattern or structure. Young students, for example, might notice that three and seven more is the same amount as seven and three more, or they may sort a collection of shapes according to how many sides the shapes have. Later, students will see 7×8 equals the well remembered $7 \times 5 + 7 \times 3$, in preparation for learning about the distributive property. In the expression $x^2 + 9x + 14$, older students can see the 14 as 2×7 and the 9 as $2 + 7$. They recognize the significance of an existing line in a geometric figure and can use the strategy of drawing an auxiliary line for solving problems. They also can step back for an overview and shift perspective. They can see complicated things, such as some algebraic expressions, as single objects or as being composed of several objects. For example, they can see $5 - 3(x - y)^2$ as 5 minus a positive number times a square and use that to realize that its value cannot be more than 5 for any real numbers x and y.

#8 Look for and express regularity in repeated reasoning.

Mathematically proficient students notice if calculations are repeated, and look both for general methods and for shortcuts. Upper elementary students might notice when dividing 25 by 11 that they are repeating the same calculations over over again, and conclude they have a repeating decimal. By paying attention to the calculation of slope as they repeatedly check whether points are on the line through (1, 2) with slope 3, middle school students might abstract the equation $(y - 2)/(x - 1) = 3$. Noticing the regularity in the way terms cancel when expanding $(x - 1)(x + 1)$, $(x - 1)(x^2 + x + 1)$, and $(x - 1)(x^3 + x^2 + x + 1)$ might lead them to the general formula for the sum of a geometric series. As they work to solve a problem, mathematically proficient students maintain oversight of the process, while attending to the details. They continually evaluate the reasonableness of their intermediate results.

CCSS Grade Level Introduction

In Grade 8, instructional time should focus on three critical areas: (1) formulating and reasoning about expressions and equations, including modeling an association in bivariate data with a linear equation, and solving linear equations and systems of linear equations; (2) grasping the concept of a function and using functions to describe quantitative relationships; (3) analyzing two- and three-dimensional space and figures using distance, angle, similarity, and congruence, and understanding and applying the Pythagorean Theorem.

1. Students use linear equations and systems of linear equations to represent, analyze, and solve a variety of problems. Students recognize equations for proportions ($y/x = m$ or $y = mx$) as special linear equations ($y = mx + b$), understanding that the constant of proportionality (m) is the slope, and the graphs are lines through the origin. They understand that the slope (m) of a line is a constant rate of change, so that if the input or x-coordinate changes by an amount A, the output or y-coordinate changes by the amount $m \cdot A$. Students also use a linear equation to describe the association between two quantities in bivariate data (such as arm span vs. height for students in a classroom). At this grade, fitting the model, and assessing its fit to the data are done informally. Interpreting the model in the context of the data requires students to express a relationship between the two quantities in question and to interpret components of the relationship (such as slope and y-intercept) in terms of the situation.

   Students strategically choose and efficiently implement procedures to solve linear equations in one variable, understanding that when they use the properties of equality and the concept of logical equivalence, they maintain the solutions of the original equation. Students solve systems of two linear equations in two variables and relate the systems to pairs of lines in the plane; these intersect, are parallel, or are the same line. Students use linear equations, systems of linear equations, linear functions, and their understanding of slope of a line to analyze situations and solve problems.

2. Students grasp the concept of a function as a rule that assigns to each input exactly one output. They understand that functions describe situations where one quantity determines another. They can translate among representations and partial representations of functions (noting that tabular and graphical representations may be partial representations), and they describe how aspects of the function are reflected in the different representations.

(continued on next page)

3. Students use ideas about distance and angles, how they behave under translations, rotations, reflections, and dilations, and ideas about congruence and similarity to describe and analyze two-dimensional figures and to solve problems. Students show that the sum of the angles in a triangle is the angle formed by a straight line, and that various configurations of lines give rise to similar triangles because of the angles created when a transversal cuts parallel lines. Students understand the statement of the Pythagorean Theorem and its converse, and can explain why the Pythagorean Theorem holds, for example, by decomposing a square in two different ways. They apply the Pythagorean Theorem to find distances between points on the coordinate plane, to find lengths, and to analyze polygons. Students complete their work on volume by solving problems involving cones, cylinders, and spheres.

The Number System

Cluster

Know that there are numbers that are not rational, and approximate them by rational numbers.

Standard

Know that numbers that are not rational are called irrational. Understand informally that every number has a decimal expansion; for rational numbers show that the decimal expansion repeats eventually, and convert a decimal expansion which repeats eventually into a rational number.

Suggested Mathematical Practices

See inside cover for correlating Enduring Understandings and Essential Questions.

#2 Reason abstractly and quantitatively.

Enduring Understandings

Rational numbers can be represented in multiple ways and are useful when examining situations involving numbers that are not whole.

Essential Questions

In what ways can rational numbers be useful?

Suggested Learning Targets

I can classify a number as rational or irrational based on its decimal expansion. (K)
I can convert a repeating decimal into a rational number. (K)

Vocabulary

rational number, irrational number

8 . NS . 1

The Number System

Cluster

Know that there are numbers that are not rational, and approximate them by rational numbers.

Standard

Use rational approximations of irrational numbers to compare the size of irrational numbers, locate them approximately on a number line diagram, and estimate the value of expressions (e.g., π^2). *For example, by truncating the decimal expansion of $\sqrt{2}$, show that $\sqrt{2}$ is between 1 and 2, then between 1.4 and 1.5, and explain how to continue on to get better approximations.*

Suggested Mathematical Practices

See inside cover for correlating Enduring Understandings and Essential Questions.

#6 Attend to precision.

Enduring Understandings	Essential Questions
Rational numbers can be represented in multiple ways and are useful when examining situations involving numbers that are not whole.	In what ways can rational numbers be useful?

Suggested Learning Targets

I can use reasoning to determine between which two consecutive whole numbers a square root will fall (e.g., I can reason that $\sqrt{39}$ is 6 and 7, because it is between $\sqrt{36}$ and $\sqrt{49}$.). (S)

I can plot the estimated value of an irrational number on a number line. (K)

I can estimate the value of an irrational number by rounding to a specific place value. (K)

I can use estimated values to compare two or more irrational numbers. (S)

Vocabulary

rational number, irrational number

8.NS.2

Expressions and Equations

Cluster

Work with radicals and integer exponents.

Standard

Know and apply the properties of integer exponents to generate equivalent numerical expressions. *For example, $3^2 \times 3^{-5} = 3^{-3} = 1/3^3 = 1/27$.*

Suggested Mathematical Practices

See inside cover for correlating Enduring Understandings and Essential Questions.

#7 Look for and make use of structure.

Enduring Understandings	Essential Questions
Algebraic expressions and equations are used to model real-life problems and represent quantitative relationships, so that the numbers and symbols can be mindfully manipulated to reach a solution or make sense of the quantitative relationships.	How can algebraic expressions and equations be used to model, analyze, and solve mathematical situations?

Suggested Learning Targets

I can determine the properties of integer exponents by exploring patterns and applying my understanding of properties of whole number exponents. (R)

I can use the properties of integer exponents to simplify expressions. (R)

Vocabulary

integer, exponent

8.EE.1

Expressions and Equations

Cluster

Work with radicals and integer exponents.

Standard

Use square root and cube root symbols to represent solutions to equations of the form $x^2 = p$ and $x^3 = p$, where p is a positive rational number. Evaluate square roots of small perfect squares and cube roots of small perfect cubes. Know that $\sqrt{2}$ is irrational.

Suggested Mathematical Practices

See inside cover for correlating Enduring Understandings and Essential Questions.

#2 Reason abstractly and quantitatively.

Enduring Understandings	Essential Questions
Algebraic expressions and equations are used to model real-life problems and represent quantitative relationships, so that the numbers and symbols can be mindfully manipulated to reach a solution or make sense of the quantitative relationships.	How can algebraic expressions and equations be used to model, analyze, and solve mathematical situations?

Suggested Learning Targets

I can recognize taking a square root as the inverse of squaring a number. (K)
I can recognize taking a cube root as the inverse of cubing a number. (K)
I can evaluate the square root of a perfect square. (S)
I can evaluate the cube root of a perfect cube. (S)
I can justify that the square root of a non-perfect square will be irrational. (R)

Vocabulary

cube, square, cube root, square root, radical, perfect square, perfect cube, irrational

8.EE.2

Expressions and Equations

Cluster

Work with radicals and integer exponents.

Standard

Use numbers expressed in the form of a single digit times a whole-number power of 10 to estimate very large or very small quantities, and to express how many times as much one is than the other. *For example, estimate the population of the United States as 3 times 10^8 and the population of the world as 7 times 10^9, and determine that the world population is more than 20 times larger.*

Suggested Mathematical Practices

See inside cover for correlating Enduring Understandings and Essential Questions.

#7 Look for and make use of structure.

Enduring Understandings	Essential Questions
Algebraic expressions and equations are used to model real-life problems and represent quantitative relationships, so that the numbers and symbols can be mindfully manipulated to reach a solution or make sense of the quantitative relationships.	How can algebraic expressions and equations be used to model, analyze, and solve mathematical situations?

Suggested Learning Targets

I can write an estimation of a large quantity by expressing it as the product of a single-digit number and a positive power of ten. (S)

I can write an estimation of a very small quantity by expressing it as the product of a single-digit number and a negative power of ten. (S)

I can compare quantities written as the product of a single-digit number and a power of ten by stating their multiplicative relationships. (R)

Vocabulary

power of ten

8 . EE . 3

Expressions and Equations

Cluster

Work with radicals and integer exponents.

Standard

Perform operations with numbers expressed in scientific notation, including problems where both decimal and scientific notation are used. Use scientific notation and choose units of appropriate size for measurements of very large or very small quantities (e.g., use millimeters per year for seafloor spreading). Interpret scientific notation that has been generated by technology.

Suggested Mathematical Practices

See inside cover for correlating Enduring Understandings and Essential Questions.

#6 Attend to precision.

Enduring Understandings	Essential Questions
Algebraic expressions and equations are used to model real-life problems and represent quantitative relationships, so that the numbers and symbols can be mindfully manipulated to reach a solution or make sense of the quantitative relationships.	How can algebraic expressions and equations be used to model, analyze, and solve mathematical situations?

Suggested Learning Targets

I can add and subtract two numbers written in scientific notation. (S)
I can multiply and divide two numbers written in scientific notation. (S)
I can select the appropriate units for measuring derived measurements when comparing quantities written in scientific notation. (S)
I can identify and interpret the various ways scientific notation is displayed on calculators and through computer software. (K)

Vocabulary

scientific notation, power of ten

8.EE.4

Expressions and Equations

Cluster

Understand the connections between proportional relationships, lines, and linear equations.

Standard

Graph proportional relationships, interpreting the unit rate as the slope of the graph. Compare two different proportional relationships represented in different ways. *For example, compare a distance-time graph to a distance-time equation to determine which of two moving objects has greater speed.*

Suggested Mathematical Practices

See inside cover for correlating Enduring Understandings and Essential Questions.

#8 Look for and express regularity in repeated reasoning.

Enduring Understandings	Essential Questions
Algebraic expressions and equations are used to model real-life problems and represent quantitative relationships, so that the numbers and symbols can be mindfully manipulated to reach a solution or make sense of the quantitative relationships.	How can algebraic expressions and equations be used to model, analyze, and solve mathematical situations?

Suggested Learning Targets

I can graph a proportional relationship in the coordinate plane. (K)

I can interpret the unit rate of a proportional relationship as the slope of the graph. (R)

I can justify that the graph of a proportional relationship will always intersect the origin $(0, 0)$ of the graph. (R)

I can use a graph, a table, or an equation to determine the unit rate of a proportional relationship and use the unit rate to make comparisons between various proportional relationships. (R)

Vocabulary

proportional relationship, unit rate, slope

8.EE.5

Expressions and Equations

Cluster

Understand the connections between proportional relationships, lines, and linear equations.

Standard

Use similar triangles to explain why the slope m is the same between any two distinct points on a non-vertical line in the coordinate plane; derive the equation $y = mx$ for a line through the origin and the equation $y = mx + b$ for a line intercepting the vertical axis at b.

Suggested Mathematical Practices

See inside cover for correlating Enduring Understandings and Essential Questions.

#1 Make sense of problems and persevere in solving them.

Enduring Understandings	Essential Questions
Algebraic expressions and equations are used to model real-life problems and represent quantitative relationships, so that the numbers and symbols can be mindfully manipulated to reach a solution or make sense of the quantitative relationships.	How can algebraic expressions and equations be used to model, analyze, and solve mathematical situations?

Suggested Learning Targets

- I can create right triangles by drawing a horizontal line segment and a vertical line segment from any two points on a non-vertical line in the coordinate plane. (S)
- I can justify that these right triangles are similar by comparing the ratios of the lengths of the corresponding legs. (R)
- I can justify that since the triangles are similar, the ratios of all corresponding hypotenuses, representing the slope of the line, will be equivalent. (R)
- I can justify that an equation in the form $y = mx$ will represent the graph of a proportional relationship with a slope of m and a y-intercept of 0. (R)
- I can justify that an equation in the form $y = mx + b$ represents the graph of a linear relationship with a slope of m and a y-intercept of b. (R)

Vocabulary

right triangle, leg, hypotenuse, similar triangles, ratio, slope, proportional relationship, y-intercept

8.EE.6

Expressions and Equations

Cluster

Analyze and solve linear equations and pairs of simultaneous linear equations.

Standard

Solve linear equations in one variable.

a. Give examples of linear equations in one variable with one solution, infinitely many solutions, or no solutions. Show which of these possibilities is the case by successively transforming the given equation into simpler forms, until an equivalent equation of the form $x = a$, $a = a$, or $a = b$ results (where a and b are different numbers).

b. Solve linear equations with rational number coefficients, including equations whose solutions require expanding expressions using the distributive property and collecting like terms.

Suggested Mathematical Practices

See inside cover for correlating Enduring Understandings and Essential Questions.

#7 Look for and make use of structure.

Enduring Understandings	Essential Questions
Algebraic expressions and equations are used to model real-life problems and represent quantitative relationships, so that the numbers and symbols can be mindfully manipulated to reach a solution or make sense of the quantitative relationships.	How can algebraic expressions and equations be used to model, analyze, and solve mathematical situations?

Suggested Learning Targets

I can use the properties of real numbers to determine the solution of a linear equation. (R)

I can simplify a linear equation by using the distributive property and/or combining like terms. (R)

I can give examples of linear equations with one solution, infinitely many solutions, or no solution. (R)

Vocabulary

linear equation, equivalent equations, rational number, coefficient, like terms, solution

8.EE.7

■ ■ ■ ■ ■ ■ ■ ■ ■ ■ ■ ■ ■ ■ ■ ■ ■

Expressions and Equations

Cluster

Analyze and solve linear equations and pairs of simultaneous linear equations.

Standard

Analyze and solve pairs of simultaneous linear equations.

a.   Understand that solutions to a system of two linear equations in two variables correspond to points of intersection of their graphs, because points of intersection satisfy both equations simultaneously.

b.   Solve systems of two linear equations in two variables algebraically, and estimate solutions by graphing the equations. Solve simple cases by inspection. *For example, $3x + 2y = 5$ and $3x + 2y = 6$ have no solution because $3x + 2y$ cannot simultaneously be 5 and 6.*

c.   Solve real-world and mathematical problems leading to two linear equations in two variables. *For example, given coordinates for two pairs of points, determine whether the line through the first pair of points intersects the line through the second pair.*

Suggested Mathematical Practices

See inside cover for correlating Enduring Understandings and Essential Questions.

#2 Reason abstractly and quantitatively.

Enduring Understandings	Essential Questions
Algebraic expressions and equations are used to model real-life problems and represent quantitative relationships, so that the numbers and symbols can be mindfully manipulated to reach a solution or make sense of the quantitative relationships.	How can algebraic expressions and equations be used to model, analyze, and solve mathematical situations?

Suggested Learning Targets

I can explain how a line represents the infinite number of solutions to a linear equation with two variables. (R)

(continued on next page)

Vocabulary

linear equation, system of linear equations (also, simultaneous linear equations), intersection

8.EE.8

Expressions and Equations

Cluster
Analyze and solve linear equations and pairs of simultaneous linear equations.

Standard
Analyze and solve pairs of simultaneous linear equations.

a. Understand that solutions to a system of two linear equations in two variables correspond to points of intersection of their graphs, because points of intersection satisfy both equations simultaneously.

b. Solve systems of two linear equations in two variables algebraically, and estimate solutions by graphing the equations. Solve simple cases by inspection. *For example, $3x + 2y = 5$ and $3x + 2y = 6$ have no solution because $3x + 2y$ cannot simultaneously be 5 and 6.*

c. Solve real-world and mathematical problems leading to two linear equations in two variables. *For example, given coordinates for two pairs of points, determine whether the line through the first pair of points intersects the line through the second pair.*

Suggested Mathematical Practices
See inside cover for correlating Enduring Understandings and Essential Questions.

#2 Reason abstractly and quantitatively.

Suggested Learning Targets

(continued from previous page)

I can explain how the point(s) of intersection of two graphs will represent the solution to the system of two linear equations because that/those point(s) are solutions to both equations.(R)

I can use algebraic reasoning (simple substitution) and the properties of real numbers to solve a system of linear equations. (S)

I can use the graphs of two linear equations to estimate the solution of the system. (S)

I can use mathematical reasoning to solve simple systems of linear equations. (S)

I can solve real-world problems and mathematical problems dealing with systems of linear equations and interpret the solution in the context of the problem. (S)

8.EE.8 *(cont.)*

Functions

Cluster

Define, evaluate, and compare functions.

Standard

Understand that a function is a rule that assigns to each input exactly one output. The graph of a function is the set of ordered pairs consisting of an input and the corresponding output.*

Function notation is not required in Grade 8.

Suggested Mathematical Practices

See inside cover for correlating Enduring Understandings and Essential Questions.

#3 Construct viable arguments and critique the reasoning of others.

Enduring Understandings

The characteristics of functions and their representations are useful in making sense of patterns and solving problems involving quantitative relationships.

Essential Questions

How are functions useful?

Suggested Learning Targets

I can explain that a function represents a relationship between an input and an output where the output depends on the input; therefore, there can be only one output for each input. (R)

I can show the relationship between the inputs and outputs of a function by graphing them as ordered pairs on a coordinate grid. (S)

Vocabulary

function, input, output

8.F.1

Functions

Cluster

Define, evaluate, and compare functions.

Standard

Compare properties of two functions each represented in a different way (algebraically, graphically, numerically in tables, or by verbal descriptions). *For example, given a linear function represented by a table of values and a linear function represented by an algebraic expression, determine which function has the greater rate of change.*

Suggested Mathematical Practices

See inside cover for correlating Enduring Understandings and Essential Questions.

#7 Look for and make use of structure.

Enduring Understandings	Essential Questions
The characteristics of functions and their representations are useful in making sense of patterns and solving problems involving quantitative relationships.	How are functions useful?

Suggested Learning Targets

I can determine the properties of a function written in algebraic form (e.g., rate of change, meaning of *y*-intercept, linear, non-linear). (S)

I can determine the properties of a function when given the inputs and outputs in a table. (S)

I can determine the properties of a function represented as a graph. (S)

I can determine the properties of a function when given the situation verbally. (S)

I can compare the properties of two functions that are represented differently (e.g., as an equation, in a table, graphically or a verbal representation). (R)

Vocabulary

function, linear function, rate of change

8.F.2

Functions

Cluster

Define, evaluate, and compare functions.

Standard

Interpret the equation $y = mx + b$ as defining a linear function, whose graph is a straight line; give examples of functions that are not linear. *For example, the function $A = s^2$ giving the area of a square as a function of its side length is not linear because its graph contains the points (1,1), (2,4) and (3,9), which are not on a straight line.*

Suggested Mathematical Practices

See inside cover for correlating Enduring Understandings and Essential Questions.

#3 Construct viable arguments and critique the reasoning of others.

Enduring Understandings	Essential Questions
The characteristics of functions and their representations are useful in making sense of patterns and solving problems involving quantitative relationships.	How are functions useful?

Suggested Learning Targets

I can explain why the equation $y = mx + b$ represents a linear function and interpret the slope and y-intercept in relation to the function. (R)

I can give examples of relationships that are non-linear functions. (K)

I can analyze the rate of change between input and output values to determine if function is linear or non-linear. (S)

I can create a table of values that can be defined as a non-linear function. (S)

Vocabulary

linear function

8.F.3

Functions

Cluster

Use functions to model relationships between quantities.

Standard

Construct a function to model a linear relationship between two quantities. Determine the rate of change and initial value of the function from a description of a relationship or from two (x, y) values, including reading these from a table or from a graph. Interpret the rate of change and initial value of a linear function in terms of the situation it models, and in terms of its graph or a table of values.

Suggested Mathematical Practices

See inside cover for correlating Enduring Understandings and Essential Questions.

#1 Make sense of problems and persevere in solving them.

Enduring Understandings	Essential Questions
The characteristics of functions and their representations are useful in making sense of patterns and solving problems involving quantitative relationships.	How are functions useful?

Suggested Learning Targets

I can write a linear function that models a given situation given verbally as a table of x- and y- values or as a graph. (S)
I can define the initial value of the function in relation to the situation. (S)
I can define the rate of change in relation to the situation. (S)
I can define the y-intercept in relation to the situation. (S)
I can explain any constraints on the domain in relation to the situation. (S)

Vocabulary

linear function, rate of change

8.F.4

Functions

Cluster

Use functions to model relationships between quantities.

Standard

Describe qualitatively the functional relationship between two quantities by analyzing a graph (e.g., where the function is increasing or decreasing, linear or nonlinear). Sketch a graph that exhibits the qualitative features of a function that has been described verbally.

Suggested Mathematical Practices

See inside cover for correlating Enduring Understandings and Essential Questions.

#4 Model with mathematics.

Enduring Understandings	Essential Questions
The characteristics of functions and their representations are useful in making sense of patterns and solving problems involving quantitative relationships.	How are functions useful?

Suggested Learning Targets

I can match the graph of function to a given situation. (S)

I can write a story that describes the functional relationship between two variables depicted on a graph. (P)

I can create a graph of function that describes the relationship between two variables. (S)

Vocabulary

increasing, decreasing, linear, nonlinear

8.F.5

Geometry

Cluster

Understand congruence and similarity using physical models, transparencies, or geometry software.

Standard

Verify experimentally the properties of rotations, reflections, and translations:

a.   Lines are taken to lines, and line segments to line segments of the same length.
b.   Angles are taken to angles of the same measure.
c.   Parallel lines are taken to parallel lines.

Suggested Mathematical Practices

See inside cover for correlating Enduring Understandings and Essential Questions.

#3 Construct viable arguments and critique the reasoning of others.

Enduring Understandings	Essential Questions
Geometric attributes (such as shapes, lines, angles, figures, and planes) provide descriptive information about an object's properties and position in space and support visualization and problem solving.	How does geometry better describe objects?

Suggested Learning Targets

- I can verify—by measuring and comparing lengths, angle measures, and parallelism of a figure and its image—that after a figure has been translated, corresponding lines and line segments remain the same length, corresponding angles have the same measure, and corresponding parallel lines remain parallel. (K)
- I can verify—by measuring and comparing lengths, angle measures, and parallelism of a figure and its image—that after a figure has been reflected, corresponding lines and line segments remain the same length, corresponding angles have the same measure, and corresponding parallel lines remain parallel. (K)
- I can verify—by measuring and comparing lengths, angle measures, and parallelism of a figure and its image—that after a figure has been rotated, corresponding lines and line segments remain the same length, corresponding angles have the same measure, and corresponding parallel lines remain parallel. (K)

Vocabulary

transformation, translation, reflection, rotation, parallel line

8.G.1

Geometry

Cluster

Understand congruence and similarity using physical models, transparencies, or geometry software.

Standard

Understand that a two-dimensional figure is congruent to another if the second can be obtained from the first by a sequence of rotations, reflections, and translations; given two congruent figures, describe a sequence that exhibits the congruence between them.

Suggested Mathematical Practices

See inside cover for correlating Enduring Understandings and Essential Questions.
#4 Model with mathematics.

Enduring Understandings

Geometric attributes (such as shapes, lines, angles, figures, and planes) provide descriptive information about an object's properties and position in space and support visualization and problem solving.

Essential Questions

How does geometry better describe objects?

Suggested Learning Targets

I can explain how transformations can be used to prove that two figures are congruent. (R)

I can perform a series of transformations (reflections, rotations, and/or translations) to prove or disprove that two given figures are congruent. (S)

Vocabulary

congruent, transformation, reflection, rotation, translation

8.G.2

Geometry

Cluster

Understand congruence and similarity using physical models, transparencies, or geometry software.

Standard

Describe the effect of dilations, translations, rotations, and reflections on two-dimensional figures using coordinates.

Suggested Mathematical Practices

See inside cover for correlating Enduring Understandings and Essential Questions.

#2 Reason abstractly and quantitatively.

Enduring Understandings	Essential Questions
Geometric attributes (such as shapes, lines, angles, figures, and planes) provide descriptive information about an object's properties and position in space and support visualization and problem solving.	How does geometry better describe objects?

Suggested Learning Targets

I can describe the changes occurring to the x- and y- coordinates of a figure after a translation. (R)

I can describe the changes occurring to the x- and y- coordinates of a figure after a reflection. (R)

I can describe the changes occurring to the x- and y- coordinates of a figure after a rotation. (R)

I can describe the changes occurring to the x- and y- coordinates of a figure after a dilation. (R)

Vocabulary

transformation, translation, reflection, rotation, dilation

8.G.3

Geometry

Cluster

Understand congruence and similarity using physical models, transparencies, or geometry software.

Standard

Understand that a two-dimensional figure is similar to another if the second can be obtained from the first by a sequence of rotations, reflections, translations, and dilations; given two similar two-dimensional figures, describe a sequence that exhibits the similarity between them.

Suggested Mathematical Practices

See inside cover for correlating Enduring Understandings and Essential Questions.

#4 Model with mathematics.

Enduring Understandings	Essential Questions
Geometric attributes (such as shapes, lines, angles, figures, and planes) provide descriptive information about an object's properties and position in space and support visualization and problem solving.	How does geometry better describe objects?

Suggested Learning Targets

I can explain how transformations can be used to prove that two figures are similar. (R)

I can describe a sequence of transformations to prove or disprove that two given figures are similar. (S)

Vocabulary

similar, transformation, reflection, rotation, translation, dilation

8.G.4

Geometry

Cluster

Understand congruence and similarity using physical models, transparencies, or geometry software.

Standard

Use informal arguments to establish facts about the angle sum and exterior angle of triangles, about the angles created when parallel lines are cut by a transversal, and the angle-angle criterion for similarity of triangles. *For example, arrange three copies of the same triangle so that the sum of the three angles appears to form a line, and give an argument in terms of transversals why this is so.*

Suggested Mathematical Practices

See inside cover for correlating Enduring Understandings and Essential Questions.

#4 Model with mathematics.

Enduring Understandings	Essential Questions
Geometric attributes (such as shapes, lines, angles, figures, and planes) provide descriptive information about an object's properties and position in space and support visualization and problem solving.	How does geometry better describe objects?

Suggested Learning Targets

I can informally prove that the sum of any triangle's interior angles will have the same measure as a straight angle (i.e., by tearing off the three corners of a triangle and arranging them to form a 180° straight angle). (R)

I can informally prove that the sum of any polygon's exterior angles will be 360-degrees. (R)

I can make conjectures regarding the relationships and measurements of the angles created when two parallel lines are cut by a transversal. (R)

I can apply proven relationships to establish minimal properties to justify similarity. (R)

Vocabulary

interior angle, exterior angle, parallel lines, transversal, similar

8.G.5

Geometry

Cluster

Understand and apply the Pythagorean Theorem.

Standard

Explain a proof of the Pythagorean Theorem and its converse.

Suggested Mathematical Practices

See inside cover for correlating Enduring Understandings and Essential Questions.

#4 Model with mathematics.
#7 Look for and make use of structure.

Enduring Understandings	Essential Questions
Geometric attributes (such as shapes, lines, angles, figures, and planes) provide descriptive information about an object's properties and position in space and support visualization and problem solving.	How does geometry better describe objects?

Suggested Learning Targets

I can use visual models to demonstrate the relationship of the three side lengths of any right triangle. (R)

I can use algebraic reasoning to relate the visual model to the Pythagorean Theorem. (R)

I can use the Pythagorean Theorem to determine if a given triangle is a right triangle. (R)

Vocabulary

Pythagorean Theorem, leg, hypotenuse, converse

8.G.6

Geometry

Cluster

Understand and apply the Pythagorean Theorem.

Standard

Apply the Pythagorean Theorem to determine unknown side lengths in right triangles in real-world and mathematical problems in two and three dimensions.

Suggested Mathematical Practices

See inside cover for correlating Enduring Understandings and Essential Questions.

#4 Model with mathematics.

Enduring Understandings

Geometric attributes (such as shapes, lines, angles, figures, and planes) provide descriptive information about an object's properties and position in space and support visualization and problem solving.

Essential Questions

How does geometry better describe objects?

Suggested Learning Targets

I can apply the Pythagorean Theorem to find an unknown side length of a right triangle. (S)

I can draw a diagram and use the Pythagorean Theorem to solve real-world problems involving right triangles. (S)

I can draw a diagram to find right triangles in a three-dimensional figure and use the Pythagorean Theorem to calculate various dimensions. (S)

Vocabulary

Pythagorean Theorem, leg, hypotenuse

8.G.7

Geometry

Cluster
Understand and apply the Pythagorean Theorem.

Standard
Apply the Pythagorean Theorem to find the distance between two points in a coordinate system.

Suggested Mathematical Practices
See inside cover for correlating Enduring Understandings and Essential Questions.

#2 Reason abstractly and quantitatively.

Enduring Understandings	Essential Questions
Geometric attributes (such as shapes, lines, angles, figures, and planes) provide descriptive information about an object's properties and position in space and support visualization and problem solving.	How does geometry better describe objects?

Suggested Learning Targets

I can connect any two points on a coordinate grid to a third point so that the three points form a right triangle. (K)

I can use the right triangle and the Pythagorean Theorem to find the distance between the original two points. (S)

Vocabulary

Pythagorean Theorem, leg, hypotenuse

8.G.8

Geometry

Cluster

Solve real-world and mathematical problems involving volume of cylinders, cones, and spheres.

Standard

Know the formulas for the volumes of cones, cylinders, and spheres and use them to solve real-world and mathematical problems.

Suggested Mathematical Practices

See inside cover for correlating Enduring Understandings and Essential Questions.
#4 Model with mathematics.

Enduring Understandings

Geometric attributes (such as shapes, lines, angles, figures, and planes) provide descriptive information about an object's properties and position in space and support visualization and problem solving.

Essential Questions

How does geometry better describe objects?

Suggested Learning Targets

I can describe the similarity between finding the volume of a cylinder and the volume of a right prism. (R)
I can recall the formula to find the volume of a cylinder. (K)
I can informally prove the relationship between the volume of a cylinder and the volume of a cone with the same base. (R)
I can recall the formula to find the volume of a cone. (K)
I can informally prove the relationship between the volume of a sphere and the volume of a circumscribed cylinder. (R)
I can recall the formula to find the volume of a sphere. (K)
I can use the formulas to find the volume of cylinders, cones, and spheres. (S)
I can solve real-world problems involving the volume of cylinders, cones, and spheres. (S)

Vocabulary

cylinder, cone, sphere, volume

8.G.9

Statistics and Probability

Cluster

Investigate patterns of association in bivariate data.

Standard

Construct and interpret scatter plots for bivariate measurement data to investigate patterns of association between two quantities. Describe patterns such as clustering, outliers, positive or negative association, linear association, and nonlinear association.

Suggested Mathematical Practices

See inside cover for correlating Enduring Understandings and Essential Questions.

#4 Model with mathematics.

Enduring Understandings	Essential Questions
The rules of probability can lead to more valid and reliable predictions about the likelihood of an event occurring.	How is probability used to make informed decisions about uncertain events?

Suggested Learning Targets

I can plot ordered pairs on a coordinate grid representing the relationship between two data sets. (K)

I can describe patterns in the plotted points such as clustering, outliers, positive or negative association, and linear or nonlinear association and describe the pattern in the context of the measurement data. (R)

I can interpret the patterns of association in the context of the data sample. (R)

Vocabulary

scatter plot, bivariate, clustering, outliers, positive association, negative association, linear association, nonlinear association

8 . SP . 1

Statistics and Probability

Cluster

Investigate patterns of association in bivariate data.

Standard

Know that straight lines are widely used to model relationships between two quantitative variables. For scatter plots that suggest a linear association, informally fit a straight line, and informally assess the model fit by judging the closeness of the data points to the line.

Suggested Mathematical Practices

See inside cover for correlating Enduring Understandings and Essential Questions.

#3 Construct viable arguments and critique the reasoning of others.

Enduring Understandings	Essential Questions
The rules of probability can lead to more valid and reliable predictions about the likelihood of an event occurring.	How is probability used to make informed decisions about uncertain events?

Suggested Learning Targets

I can recognize whether or not data plotted on a scatter plot have a linear association. (K)

I can draw a straight trend line to approximate the linear relationship between the plotted points of two data sets. (S)

I can make inferences regarding the reliability of the trend line by noting the closeness of the data points to the line. (R)

Vocabulary

scatter plot, linear association, trend line, line of best fit

Statistics and Probability

Cluster

Investigate patterns of association in bivariate data.

Standard

Use the equation of a linear model to solve problems in the context of bivariate measurement data, interpreting the slope and intercept. *For example, in a linear model for a biology experiment, interpret a slope of 1.5 cm/hr as meaning that an additional hour of sunlight each day is associated with an additional 1.5 cm in mature plant height.*

Suggested Mathematical Practices

See inside cover for correlating Enduring Understandings and Essential Questions.

#2 Reason abstractly and quantitatively.

Enduring Understandings	Essential Questions
The rules of probability can lead to more valid and reliable predictions about the likelihood of an event occurring.	How is probability used to make informed decisions about uncertain events?

Suggested Learning Targets

I can determine the equation of the trend line that approximates the linear relationship between the plotted points of two data sets. (S)

I can interpret the *y*-intercept of the equation in the context of the collected data. (R)

I can interpret the slope of the equation in the context of the collected data. (R)

I can use the equation of the trend line to summarize the given data and make predictions regarding additional data points. (R)

Vocabulary

linear model, bivariate, slope, *y*-intercept, trend line, line of best fit

8.SP.3

Statistics and Probability

Cluster

Investigate patterns of association in bivariate data.

Standard

Understand that patterns of association can also be seen in bivariate categorical data by displaying frequencies and relative frequencies in a two-way table. Construct and interpret a two-way table summarizing data on two categorical variables collected from the same subjects. Use relative frequencies calculated for rows or columns to describe possible association between the two variables. *For example, collect data from students in your class on whether or not they have a curfew on school nights and whether or not they have assigned chores at home. Is there evidence that those who have a curfew also tend to have chores?*

Suggested Mathematical Practices

See inside cover for correlating Enduring Understandings and Essential Questions.

#3 Construct viable arguments and critique the reasoning of others.

Enduring Understandings	Essential Questions
The rules of probability can lead to more valid and reliable predictions about the likelihood of an event occurring.	How is probability used to make informed decisions about uncertain events?

Suggested Learning Targets

I can create a two-way table to record the frequencies of bivariate categorical values. (K)

I can determine the relative frequencies for rows and/or columns of a two-way table. (S)

I can use the relative frequencies and context of the problem to describe possible associations between the two sets of data. (R)

Vocabulary

bivariate, categorical data, two-way table, frequency, relative frequency

8.SP.4